Roaring ROCKETS

For Danny Spiegelhalter and Virgil Tracy – T.M.

KINGFISHER
An imprint of Kingfisher Publications Plc
New Penderel House, 283-288 High Holborn, London WC1V 7HZ
www.kingfisherpub.com

First published by Kingfisher 1997
This edition first published by Kingfisher 2007
2 4 6 8 10 9 7 5 3 1

A CIP catalogue record for this book is available from the British Library.

ISBN 978 0 7534 1597 9

Printed in China
1TR/0507/TIMS/FR(SCHOY)/157MA/C

Roaring ROCKETS

Tony Mitton
and
Ant Parker

KINGFISHER

Rockets have power. They rise and roar.

This rocket's waiting, ready to soar.

Rockets carry astronauts with cool, white suits,

oxygen helmets and gravity boots.

The countdown is finishing: 3, 2, 1 . . .

Action! Blast off! The journey's begun.

Rockets have fuel in great big tanks.

When they're empty, they drop away...thanks!

Up in space you're really light,

so astronauts strap themselves in tight.

Rockets explore. Through space they zoom

reaching as far as the big, round moon.

Out comes the lander with legs out ready

and fiery boosters to hold it steady.

Rockets take astronauts out to a place

that's strange and wonderful: silent space…

Moon mission over, the lander's left outside.

We're back in the rocket for the long return ride.

Rockets re-enter in a fiery flash
to land at sea with a sizzling splash!

The helicopter carries the brave crew away.
Let's give them a cheer. Hip, hip, hooray!

Rocket bits

gravity boots

gravity keeps us on the ground but there is not a lot on the moon so boots are worn that grip the ground and stop you floating off

lunar lander

this takes astronauts down from the rocket to land on the moon

oxygen helmet

we need to breathe oxygen but there is none in space, so astronauts carry their own supply whic flows into their helmets

fuel tanks

command module

these hold the fuel which makes the rocket go

This is the part of the rocket in which astronauts travel on their way to the moor